Mathematical Methods in Nonlinear Heat Transfer:
A Semi-Analytical Approach

Mathematical Methods in Nonlinear Heat Transfer: A Semi-Analytical Approach

Davood Domairry Ganji

&

Ehsan Mohseni Languri

Library of Congress Control Number: 2010916348
ISBN: Hardcover   978-1-4568-0829-7
               Softcover   978-1-4568-0828-0
               Ebook   978-1-4568-0830-3

This book was printed in the United States of America.

Affiliation for Authors:

Davood Domairry Ganji:
Department of Mechanical Engineering,
Noshirvani Technical University of Babol, Iran
AND
Ehsan Mohseni Languri:
Department of Mechanical Engineering,
University of Wisconsin-Milwaukee, USA

To order additional copies of this book, contact:
Xlibris Corporation
1-888-795-4274
www.Xlibris.com
Orders@Xlibris.com
86425

CONTENTS

LIST OF PICTURES
AND TABLES

CHAPTER 1

1. Introduction to Nonlinear Partial Differential Equations

Fluid flow and heat transfer exist in our daily life as well as usual industrial problems. Simple examples of such common phenomena are the flowing water in a riverbed, flying airplanes above the earth, boiling water for making tea, etc. Such simple phenomena can be modeled mathematically by means of Partial Differential Equations (PDEs). Most of the natural PDEs in the life are nonlinear due to the complexity of the problem and the number of variables involved in the problem. There are several ways to tackle such problems—such as numerical methods, experimental methods, analytical methods, and semi-Analytical methods. All these methods have their own advantages and disadvantages. Among them, the best method is the analytical method in terms of exactness and completeness. One can see the effect of different variables in the solution, which usually is not the case in other methods. The main disadvantage of the analytical method is the lack of analytical solution for all problems. Still, there are a lot of fluid flow and heat-transfer problems that do not have the exact solutions due to the complexity of the problem. In order to solve such weakness, the semi-analytical methods, including several different methods, have been introduced and used successfully over time. The main goal of the current book is to introduce such methods and their power to solve nonlinear partial differential equations. In chapters 2 to 5 of this book, four powerful semi-analytical methods are introduced and applied to different heat-transfer problems. The authors hope one can use such methods to solve complicated nonlinear partial differential equations arising in heat transfer where there is no exact solution for them (1, 2).

A partial differential equation (PDE) is an equation that includes more than one independent variables and partial derivatives of an unknown function with respect to one or more independent variable(s). The order of a PDE is defined as the order of highest derivatives involved in the equation. Solutions to partial differential equations can be divided into two main categories: particular solutions and general solutions. A function which turns the PDE into an identity when substituted in to the PDE is called a particular solution. On the other hand, a general solution contains all available particular solutions of the PDE. Usually, a solution refers to a particular solution in mathematical speaking (1, 2). A first-order partial differential equation has the following general form:

$$F(x_1, x_2, ..., x_n, \omega, \frac{\partial \omega}{\partial x_1}, \frac{\partial \omega}{\partial x_2}, ..., \frac{\partial \omega}{\partial x_n}) = 0 \qquad (1.1)$$

Here, ω is the unknown function, and F is the given function.

On the other hand, a general form of the second-order linear PDE with two independent variables (the minimum variable for this case) can be written as follows:

$$a(x,y)\frac{\partial^2 \omega}{\partial x^2} + 2b(x,y)\frac{\partial^2 \omega}{\partial x \partial y} + c(x,y)\frac{\partial^2 \omega}{\partial y^2} = \alpha(x,y)\frac{\partial \omega}{\partial x} + \beta(x,y)\frac{\partial \omega}{\partial y} + \gamma(x,y)\omega + \delta(x,y) \qquad (1.2)$$

Where functions of $a(x,y), 2b(x,y)$ $c(x,y), \alpha(x,y), + \beta(x,y), \gamma(x,y), \delta(x,y)$ are coefficients of the equation. The equation is called homogeneous of $\delta = 0$; otherwise, it is called nonhomogeneous.

In addition to linear partial differential equations, there is a category called nonlinear partial differential equations, which have the nonlinear terms in their PDEs. These nonlinear PDEs describe a wide range of more complex physical systems. The problem associated to such nonlinear partial differential equations is that there is no general technique that is applicable to all nonlinear PDEs; therefore, each of them should be studied separately. The second-order nonlinear PDE with two independent variables can be written as the following general form:

$$F(x, y, \omega, \frac{\partial \omega}{\partial x}, \frac{\partial \omega}{\partial y}, \frac{\partial^2 \omega}{\partial x^2}, \frac{\partial^2 \omega}{\partial x \partial y}, \frac{\partial^2 \omega}{\partial y^2}) = 0 \tag{1.3}$$

There is multiplication of at least two functions. Here is the list of some famous nonlinear partial differential equations encountered frequently in engineering problems (3-17):

1. **Nonlinear Heat Equation**
   The 1-D unsteady state thermal transport equation in quiescent media or solids assuming temperature dependent thermal diffusivity can be modeled by the following nonlinear PDE.

$$\frac{\partial \omega}{\partial t} = \frac{\partial}{\partial x}\left[f(\omega)\frac{\partial \omega}{\partial x} \right] \tag{1.4}$$

2. **Kolmogorov-Petrovskii-Piskunov Equation**
   This equation models problems with heat and mass transfer $(a > 0)$. This equation is also called heat equation with a nonlinear source.

$$\frac{\partial \omega}{\partial t} = a\frac{\partial^2 \omega}{\partial x^2} + f(\omega) \tag{1.5}$$

3. **Burgers Equation**
   The Burgers equation models problems in gas dynamics, showing the wave process. The general form of this equation is given below.

$$\frac{\partial \omega}{\partial t} + \omega\frac{\partial \omega}{\partial x} = \frac{\partial^2 \omega}{\partial x^2} \tag{1.6}$$

4. **Nonlinear Wave Equation**
   This type of nonlinear equation exists in wave and gas dynamics equations. Assuming $f(\omega) > 0$, we have

$$\frac{\partial^2 \omega}{\partial t^2} = \frac{\partial}{\partial x}\left[f(\omega)\frac{\partial \omega}{\partial x}\right] \tag{1.7}$$

5. Nonlinear Klein-Gorden Equation

A wide range of topics in Physics such as superconductivity, crystal dislocations, laser pulses in two-phase media, etc., are modeled through this nonlinear partial differential equation, assuming $a > 0$.

$$\frac{\partial^2 \omega}{\partial t^2} = a\frac{\partial^2}{\partial x^2} + f(\omega) \tag{1.8}$$

6. Nonlinear Laplace Equation

This famous equation in heat transfer, also called a stationary heat equation with a nonlinear source, is described below.

$$\frac{\partial^2 \omega}{\partial x^2} + \frac{\partial^2 \omega}{\partial y^2} = f(\omega) \tag{1.9}$$

7. Monge-Ampere Equation

This nonlinear partial differential equation arises in gas dynamics and meteorology with the following general form.

$$\left(\frac{\partial^2 \omega}{\partial x \partial y}\right)^2 - \frac{\partial^2 \omega}{\partial x^2}\frac{\partial^2 \omega}{\partial y^2} = f(x, y) \tag{1.10}$$

8. Navier-Stokes Equation

This famous nonlinear partial differential equation describes the motion of fluid substances applying Newton's second law to moving fluid.

$$\rho(\frac{\partial v_i}{\partial t} + v_j\frac{\partial v_i}{\partial x_j}) = -\frac{\partial p}{\partial x_i} + \frac{\partial}{\partial x_j}\left[\mu\left(\frac{\partial v_i}{\partial x_j} + \frac{\partial v_j}{\partial x_i}\right) + \lambda\frac{\partial v_k}{\partial x_k}\right] + f_i \tag{1.11}$$

There are much more nonlinear partial differential equations in a wide range of engineering applications. As mentioned earlier, there is no general rule for the analytical solution for nonlinear PDEs, and usually, the solutions are limited in some special cases. Another mathematical approach to deal with such complex equations is the approximate or semi-analytical methods. In the following chapters, some of the powerful mathematical techniques to deal with nonlinear partial differential equations that arise in heat transfer are introduced and applied.

1.1 Nomenclature

f	a function	z	a variable
t	time	ω	a function
x	a variable	∂	an operator
y	a variable	δ	a function

1.2 Reference:

1. Ioannis P Stavroulakis, Stepan A Tersian, Partial differential equations, second edition, world scientific, 2004.
2. Jurgen Jost, Graduate text in mathematics, partial differential equations, Spriner, second edition.
3. R. Courant and D. Hilbert, Methods of Mathematical Physics. Volume 2. Partial Differential Equations, Wiley-VCH, 1989.
4. L. C. Evans, Partial Differential Equations, American Mathematical Society, Providence, 1998.
5. S. J. Farlow, Partial Differential Equations for Scientists and Engineers, Dover Publications Inc., 1993.
6. 6. F. John, Partial Differential Equations. Fourth Edition, Springer, 1991.
7. J. Jost, Partial Differential Equations, Springer-Verlag, New York, 2002.
8. I. G. Petrovskii, Partial Differential Equations, W. B. Saunders Co., Philadelphia, 1967.

9. Y. Pinchover and J. Rubinstein, An Introduction to Partial Differential Equations, Cambridge University Press, Cambridge, 2005.

10. A. D. Polyanin, Handbook of Linear Partial Differential Equations for Engineers and Scientists, Chapman & Hall/CRC Press, Boca Raton, 2002.

11. A. D. Polyanin and V. F. Zaitsev, Handbook of Nonlinear Partial Differential Equations, Chapman & Hall/CRC Press, Boca Raton, 2004.

12. A. D. Polyanin, V. F. Zaitsev, and A. Moussiaux, Handbook of First Order Partial Differential Equations, Taylor & Francis, London, 2002.

13. D. L. Powers, Boundary Value Problems, Fifth Edition: and Partial Differential Equations, Elsevier Academic Press, 2005.

14. W. E. Schiesser, Computational Mathematics in Engineering and Applied Science: ODEs, DAEs, and PDEs, CRC Press, Boca Raton, 1993.

15. I. Stakgold, Boundary Value Problems of Mathematical Physics, Vols. I, II, SIAM, Philadelphia, 2000.

16. A. N. Tikhonov and A. A. Samarskii, Equations of Mathematical Physics, Dover Publ., New York, 1990.

17. D. Zwillinger, Handbook of Differential Equations (3rd edition), Academic Press, Boston, 1997.

CHAPTER 2

2. Variational Iteration Method

This section is devoted to a complete introduction of the concept of Variational Iteration Method (VIM) (1-7). To clarify the basic ideas of VIM, we consider the following differential equation:

$$Lu + Nu = g(x, y) \qquad (2.1)$$

L represents a linear operator, N a nonlinear operator, and finally, $g(x, y)$ is an inhomogeneous term. According to VIM, we can express the following correction functional in $x-$ and $y-$ directions, respectively, as follows:

$$u_{n+1}(x, y) = u_n(x, y) + \int_0^x \lambda_1 \left(Lu_n(\tau, y) + N\widetilde{u}_n(\tau, y) - g(\tau, y) \right) d\tau \qquad (2.2)$$

$$u_{n+1}(x, y) = u_n(x, y) + \int_0^y \lambda_2 \left(Lu_n(x, \tau) + N\widetilde{u}_n(x, \tau) - g(x, \tau) \right) d\tau \qquad (2.3)$$

Here, λ_1 and λ_2 are general Lagrange multipliers, which can be identified optimally by the variational theory (1-2). The subscript n

indicates the nth approximation, and $\tilde{u}_n$ is a restricted variation, which means $\delta \tilde{u}_n = 0$. Consequently, the solution is given.

$$u(x, y) = \lim_{n \to \infty} u_n (x, y) \tag{2.4}$$

2.1 Natural Convection in Porous Medium (8)

Consider an inverted cone with semiangle γ, and take axes in the manner indicated in figure 3.1 (9, 10). The boundary layer develops over the heated frustum $x = x_0$ (11, 12). One can write the boundary layer development parameters in terms of the stream function ψ as

$$u = \frac{1}{r} \frac{\partial \psi}{\partial y}, \qquad v = -\frac{1}{r} \frac{\partial \psi}{\partial x}. \tag{2.5}$$

The boundary layer equations are as follows:

$$\frac{1}{r} \frac{\partial^2 \psi}{\partial y^2} = \frac{g \beta K}{v} \frac{\partial T}{\partial y}$$

$$\frac{1}{r} \left(\frac{\partial \psi}{\partial y} \frac{\partial T}{\partial x} - \frac{\partial \psi}{\partial x} \frac{\partial T}{\partial y} \right) = \alpha_m \frac{\partial^2 T}{\partial y^2} \tag{2.6}$$

For a thin boundary layer, one can approximately $r = x \sin (\gamma)$. Suppose that either a power law of temperature or a power law of heat flux is prescribed on the frustum. Accordingly, the boundary conditions are the following:

$$
\begin{aligned}
&u = 0, T = T_\infty && y \to \infty \\
&u = 0 && y = 0, x_0 \leq x < \infty \\
&q''_m = -k_m \frac{\partial T}{\partial y}\bigg|_{y=0} = A(x - x_0)^\lambda && y = 0, x_0 \leq x < \infty
\end{aligned} \tag{2.7}
$$

There is a similar solution for the case of full cone where $x_0 = 0$, which is demonstrated in figure 1.b.

In the case of prescribed wall temperature, we let

$$\psi = \alpha_m r Ra_x^{1/3} f(g),$$

$$T - T_\infty = \frac{q_w'' x}{k_m} Ra_x^{-1/3} \theta(g), \tag{2.8}$$

$$g = \frac{y}{x} Ra_x^{1/3}.$$

where the Rayleigh number is based on heat flux.

$$Ra_x = \frac{g \beta K \cos(\gamma) q_w'' x^2}{\upsilon \alpha_m k_m} \tag{2.9}$$

The governing equation then reduces to

$$f' = h.$$
$$h'' + \left(\frac{k+5}{2}\right) f h' - \frac{2k+1}{3} f' h = 0 \tag{2.10}$$

subjected to

$$f(0) = 0, h'(0) = -1, h(\infty) = 0 \tag{2.11}$$

Final form of equations 2.10 and boundary conditions 2.11 is reduced to

$$\begin{cases} f''' + \left(\frac{k+5}{2}\right) f f'' - \left(\frac{2k+1}{3}\right)(f')^2 = 0, \\ f(0) = 0, f''(0) = -1, f'(\infty) = 0. \end{cases} \tag{2.12}$$

It is of interest to obtain the value of the local Nusselt number, which is defined as

$$Nu_x = \frac{q_w x}{k(T_w - T_\infty).}$$

(2.13)

From equations 2.8, 2.9, and 2.13, one can obtain the local Nusselt number as

$$Nu_x = Ra_x^{1/3} \left[-h(0) \right]$$

(2.14)

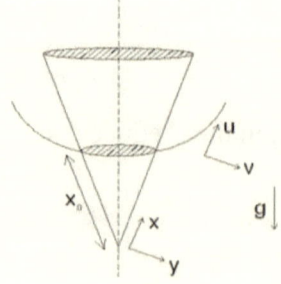

Figure 2.1: Coordinate system for a full cone ($x_0 = 0$).

To solve equation (2.12) using the VIM, we have the correction functional as

$$f(g) = 1 - e^{-g} + \int_0^g \lambda [f'''(s) + \frac{1}{2}(k+5)f(s)f''(s) - \frac{1}{3}(2k+1)(f'(s))^2 ds,$$

(2.15)

then

$$f(g) = 1 - e^{-g} + \int_0^g (s-g)[f'''(s) + \frac{1}{2}(k+5)f(s)f''(s) - \frac{1}{3}(2k+1)(f'(s))^2 ds,$$

(2.16)

so finally we have

$$f(g) = \frac{1}{24} + \frac{1}{2}e^{-g} - \frac{13}{24}k + \frac{5}{12}g + \frac{7}{12}gk + \frac{1}{2}ke^{-g} + \frac{1}{24}ke^{-2g} - \frac{13}{24}e^{-2g} \qquad (2.17)$$

and also

$$f'(g) = -\frac{1}{2}e^{-g} + \frac{5}{12} + \frac{7}{12}k - \frac{1}{2}ke^{-g} - \frac{1}{12}ke^{-2g} + \frac{13}{12}e^{-2g} . \qquad (2.18)$$

So now, for different amounts of k and g, we can compare the answers.

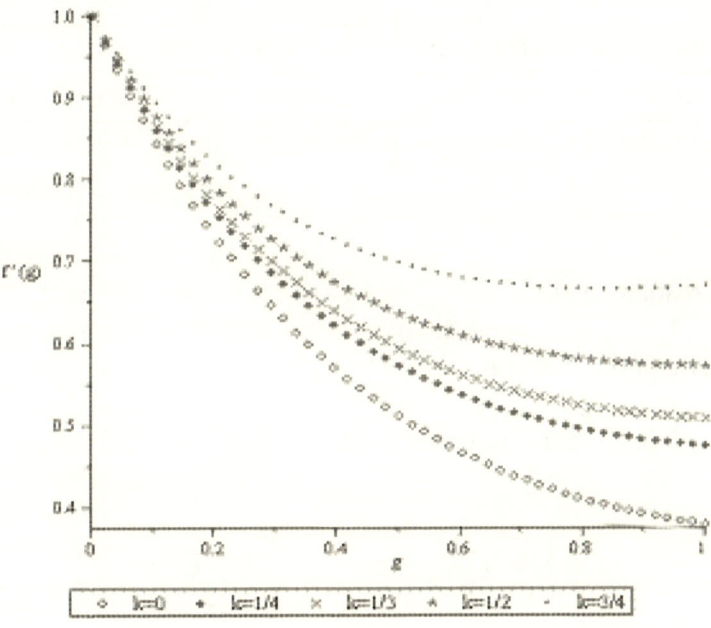

Figure 2.2: f′(g) for different amounts of K

Table 2.1: f′(g) for different amounts of k

g	K = 0	K = 1/4	K = 1/3	K = 1/2	K = 3/4
0	1.0	1.0	1.0	1.0	1.0
0.1	0.8512062735	0.8668780390	0.8721019608	0.8825498043	0.8982215697
0.2	0.7334813399	0.7630083282	0.7728506576	0.7925353164	0.8220623046
0.3	0.6408034954	0.6826009755	0.6965334688	0.7243984554	0.7661959354
0.4	0.5682796881	0.6209619956	0.6385227647	0.6736443031	0.7263266107
0.5	0.5119373982	0.5742902440	0.5950745259	0.6366430898	0.6989959357
0.6	0.4685545782	0.5395115775	0.5631639107	0.6104685769	0.6814255764
0.7	0.4355207257	0.5141434592	0.5403510371	0.5927661929	0.6713889265
0.8	0.4107234125	0.4961844478	0.5246714596	0.5816454831	0.6671065186
0.9	0.3924556324	0.4840240314	0.5145468311	0.5755924304	0.6671608295
1	0.3793401696	0.4763690876	0.5087120604	0.5733980057	0.6704269240

The results for $f'(g)$ have been shown in figure 2.2 with selected K as 0, 1/4, 1/3, 1/2, and 3/4. Also, it has been made as a comparison between the VIM solution and numerical solution in table 2.2.

Table 2.2:-f″ values obtained by Numerical and VIM solution for k = 1/4

g	K = 1/4−VIM	K = 1/4−NUM
0	1	0.911295
0.1	0.866878	0.813604
0.2	0.763008	0.721351
0.3	0.682600	0.635531
0.4	0.620961	0.556661
0.5	0.574290	0.484997

2.2 Nomenclature

A	prescribed constant	θ	similarity function for temperature
f	similarity function for stream function	η	Independent dimensionless parameter
g	acceleration due to gravity	Ψ	stream function
K	permeability of the fluid-saturated porous medium	λ	prescribed constants
Nu	local Nusselt number	β	expansion coefficient of the fluid
q_w	surface heat flux	T	temperature
r	local radius of the cone	T_∞	ambient temperature
Ra_x	local Rayleigh number	T_w	wall temperature
u, v	velocity vector along x, y axis	v	kinematic viscosity of the fluid
x, y	Cartesian coordinate system	α_m	thermal diffusivity the fluid-saturated porous medium
x_0	distance of start point of cone from the vertex		

2.3 References

1. M. Inokuti, et al., General use of the Lagrange multiplier in nonlinear mathematical physics, in: S. Nemat-Nassed (Ed), Variational Method in the Mechanics of Solids, Pergamon Press, 1978, pp. 156-162.
2. B. A. Finlayson, The Method of Weighted Residuals and Variational Principles, Academic Press, 1972.
3. Konuralp, A., the steady temperature distributions with different types of nonlinearities. (in press) (2009).

4. J. H. He, variational iteration method for autonomous ordinary differential systems, Applied Mathematics and Computation, 114(2-3). 115-123 (2000).
5. J. H. He, Approximate solution of nonlinear differential equations with convolution product nonlinearities, Computer Methods in Applied Mechanics and Engineering, 167. 69-73 (1998).
6. D. D. Ganji, A. Sadighi, Application of homotopy-perturbation and variational iteration methods to nonlinear heat transfer and porous media equations, Journal of Computational and Applied Mathematics. 207(1) 24-34(2007).
7. D. D. Ganji, M. Jannatabadi, E. Mohseni, Application of He's variational iteration method to nonlinear Jaulent–Miodek equations and comparing it with ADM, Journal of Computational and Applied Mathematics.
8. Y Vazifeshenas, D. D. Ganji, New analytical solution for natural convection of Darcian fluid in porous media prescribed surface heat flux, submitted to Communications in Nonlinear Science and Numerical Simulation, 2009.
9. K. Vafai, (ed.): Handbook of Porous Media, Marcel Dekker, New York (2000).
10. I. Pop, T. Y. Na, Natural convection over a frustum of a wavy cone in a porous medium, Mech. Res. Comm. 22.181-190 (1995).
11. D. A. Nield, A. Bejan, Convection in Porous Media, third ed., Springer-Verlag, New York, (2006).
12. P. Cheng, T. T. Le, I. Pop, Natural convection of a Darcian fluid about a cone, International Communications Heat Mass Transfer 12. 705-717 (1985).

CHAPTER 3

3. Adomian Decomposition Method

This section introduces the powerful approximate method of Adomian Decomposition (1-10). Using this method, one can easily handle nonlinear problems with the large order of nonlinearity. Let us discuss a brief outline of the Adomian Decomposition method. For this, we consider a general nonlinear equation in the form

$$Lu + Nu + Ru = g, \tag{3.1}$$

where L is the highest order derivative which is assumed to be easily invertible, R is the linear differential operator of less order than L, Nu presents the nonlinear terms, and g is the source term. Applying the inverse operator L^{-1} to the both sides of equation 3.1 and using the given conditions, we obtain

$$L^{-1}Lu = L^{-1}g - L^{-1}Ru - L^{-1}Nu. \tag{3.2}$$

If L is a second-order operator, L^{-1} will be a twofold indefinite integral. By solving equation 3.2, we have

$$u = A + Bt + L^{-1}g - L^{-1}Ru - L^{-1}Nu, \tag{3.3}$$

where A and B are constants of integration and can be found from the boundary or initial conditions. The Adomian method assumes the solution u can be expanded into an infinite series as

$$u = \sum_{m=0}^{\infty} u_m .$$ (3.4)

Also, the nonlinear term Nu will be written as

$$Nu = \sum_{m=0}^{\infty} A_m,$$ (3.5)

where A_m are the special Adomian polynomials. By specified A_m, the next component of u can be determined:

$$u_{n+1} = L^{-1} \sum_{m=0}^{m} A_m.$$ (3.6)

Finally, after some iterations and getting sufficient accuracy, the solution of the equation can be expressed by equation 3.4. In equation 3.5, the Adomian polynomials can be generated by several means. Here we used the following recursive formulation:

$$A_m = \frac{1}{m!}\left[\frac{d^m}{d\lambda^m}\left[N\left(\sum_{i=0}^{m} \lambda^i u_i\right)\right]\right]_{\lambda=0} , \quad m = 0,1,2,3,4,....$$ (3.7)

Since the method does not resort to linearization or assumption of weak nonlinearity, the solution generated is in general more realistic than those achieved by simplifying the model of the physical problem.

3.1 Natural Convection Between Two Vertical Plates (11)

A schematic of the problem (12-14) under study is shown in figure 3.1. It consists of two flat plates that can be positioned vertically.

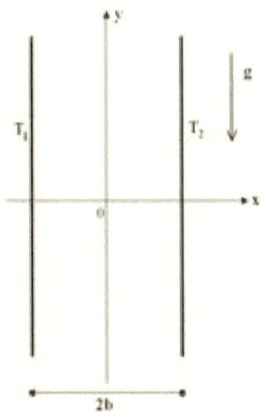

Figure 3.1 Schematic diagram of the problem under consideration

A non-Newtonian fluid is in two flat plates a distance $2b$ apart. The walls at $x = +b$ and $x = -b$ are held at constant temperatures Θ_2 and Θ_1 respectively, where $\Theta_1 > \Theta_2$. This difference in temperature causes the fluid near the wall at $x = -b$ to rise and the fluid near the wall at $x = +b$ to fall. The equation of motion is (for more details, please see ref. 14)

$$\mu \frac{d^2v}{dx^2} + 6\beta_3 (\frac{dv}{dx})^2 \frac{d^2v}{dx^2} + \rho_0 \gamma (\theta - \theta_m)g = 0, \tag{3.8}$$

And the energy equation is as follows:

$$\mu (\frac{dv}{dx})^2 + 2\beta_3 (\frac{dv}{dx})^4 + k \frac{d^2\theta}{dx^2} = 0, \tag{3.9}$$

Rajagopal (14) has demonstrated that by using the similarity variables

$$v = \frac{V}{V_0} \quad , \quad \eta = \frac{x}{b} \quad , \quad \theta = \frac{\Theta - \Theta_m}{\Theta_1 - \Theta_2}, \tag{3.10}$$

the Navier-Stokes and Energy equations can be reduced to the following

pair of ordinary differential equations (14):

$$\frac{d^2v}{d\eta^2} + 6\delta\left(\frac{dv}{d\eta}\right)^2\frac{d^2v}{d\eta^2} + \theta = 0, \tag{3.11}$$

$$\frac{d^2\theta}{d\eta^2} + E.\Pr\left(\frac{dv}{d\eta}\right)^2 + 2\delta E.\Pr\left(\frac{dv}{d\eta}\right)^4 = 0, \tag{3.12}$$

where

$$E = \frac{V_0^2}{c(\Theta_1 - \Theta_2)}, \tag{3.13}$$

$$\Pr = \frac{\mu c}{k}, \tag{3.14}$$

and

$$\delta = \frac{6\beta_3 V_0^2}{\mu b^2}. \tag{3.15}$$

where c is the specific heat of the fluid. The appropriate boundary conditions are

$$v = 0, \quad \theta = \frac{1}{2} \quad \text{at} \quad \eta = -1, \tag{3.16}$$

$$v = 0, \quad \theta = -\frac{1}{2} \quad \text{at} \quad \eta = +1, \tag{3.17}$$

We will apply the ADM to nonlinear ordinary differential equations 3.11 and 3.12. To solve these equations using ADM, we have the linear operator defined as

$$L = \frac{d^2}{d\eta^2}. \tag{3.18}$$

According to equation 3.18, equations 3.11 and 3.12 must be written as the following:

$$Lv = -\left(6\delta \left(\frac{dv}{d\eta} \right)^2 \frac{d^2v}{d\eta^2} + \theta \right) \tag{3.19}$$

$$L\theta = -\left(E.\Pr \left(\frac{dv}{d\eta} \right)^2 + 2\delta.\Pr \left(\frac{dv}{d\eta} \right)^4 \right) \tag{3.20}$$

Assume the inverse of the operators, L exists and it can be integrated from 0 to η, i.e., $L = \iint (\bullet) d\eta d\eta$. Operating with L^{-1} on equations 3.19 and 3.20 and after exerting boundary condition on them, we have

$$v(\eta) = -6\delta L^{-1}\left(N_1(v) \right) - L^{-1}\theta + C_1 + C_2\eta, \tag{3.21}$$

$$\theta(\eta) = -E.\Pr N_2(v) - 2\delta E.\Pr N_3(v) + T_1 + T_2\eta, \tag{3.22}$$

where

$$N_1(v) = \left(\frac{dv}{d\eta} \right)^2 \frac{d^2v}{d\eta^2}, \tag{3.23}$$

$$N_2(V) = \left(\frac{dv}{d\eta} \right)^2, \tag{3.24}$$

$$N_3(V) = \left(\frac{dv}{d\eta} \right)^4. \tag{3.25}$$

By using the boundary condition, we assume

$$T_1 = 0, T_2 = -\frac{1}{2}. \tag{3.26}$$

We consider

$$\theta = \sum_{n=0}^{\infty} \theta_n ,\tag{3.27}$$

$$v = \sum_{n=0}^{\infty} v_n \tag{3.28}$$

N_1, N_2, N_3 are nonlinear terms. Hence, using equation 3.19 gives

$$L^{-1}(N_1) = \sum A_n = \sum_{n=0}^{m} v''_{m-n} (\sum_{k=0}^{n} v'_{n-k} v'_k) ,\tag{3.29}$$

$$L^{-1}(N_2) = \sum B_n = \sum_{n=0}^{m} v'_{m-n} v'_n ,\tag{3.30}$$

$$L^{-1}(N_3) = \sum R_n = (\sum_{n=0}^{m} \sum_{i=0}^{n} v'_i v'_{n-i}) \sum_{j=0}^{m-n} v'_j v'_{m-n-j} ,\tag{3.31}$$

where

$$A_0 = v''_0 v'^2_0 ,\tag{3.32}$$

$$A_1 = v''_1 v'^2_0 + 2v''_0 v'_1 v'_0 ,\tag{3.33}$$

$$A_2 = v''_2 v'^2_0 + 2v''_1 v'_1 v'_0 + 2v''_0 v'_2 v'_0 + v''_0 v'^2_1 ,\tag{3.34}$$

$$\vdots$$

And

$$B_0 = v'^2_0 ,\tag{3.35}$$

$$B_1 = 2v'_0 v'_1 ,\tag{3.36}$$

$$B_2 = v'^2_0 + 2v'_0 v'_2 ,\tag{3.37}$$

$$\vdots$$

Also

$$R_0 = v_0'^4,$$ (3.38)

$$R_1 = 4v_0'^3 v_1',$$ (3.39)

$$R_2 = 4v_0'^3 v_2' + 6v_1'^2 v_0'^2,$$ (3.40)

C_1 and C_2 are obtained by checking boundary conditions in the previous step. By using boundary conditions, we have

$$v_0(\eta) = 0,$$ (3.41)

$$\theta_0(\eta) = -\frac{1}{2}\eta.$$ (3.42)

Now by using equations 3.21 and 3.22, $v(\eta)$ and $\theta(\eta)$ are obtained.

$$v_1(\eta) = \frac{1}{12}\eta^3 - \frac{1}{12}\eta$$ (3.43)

$$v_2(\eta) = 0$$ (3.44)

$$v_3(\eta) = 0$$ (3.45)

$$v_4(\eta) = -3\delta\left(\frac{1}{672}\eta^7 - \frac{1}{480}\eta^5 + \frac{1}{864}\eta^3\right) + E\,\text{Pr}\left(\frac{1}{3456}\eta^4 + \frac{1}{26880}\eta^8 - \frac{1}{8640}\eta^6\right)$$
$$+ \frac{17}{10080}\delta\eta - \frac{17}{80640}E\,\text{Pr}$$ (3.46)

$\vdots$

$$\theta_1(\eta) = 0$$ (3.47)

$$\theta_2(\eta) = 0$$ (3.48)

$$\theta_3(\eta) = -E\,\Pr\left(\frac{1}{288}\eta^2 + \frac{1}{480}\eta^6 - \frac{1}{288}\eta^4\right)$$ (3.49)

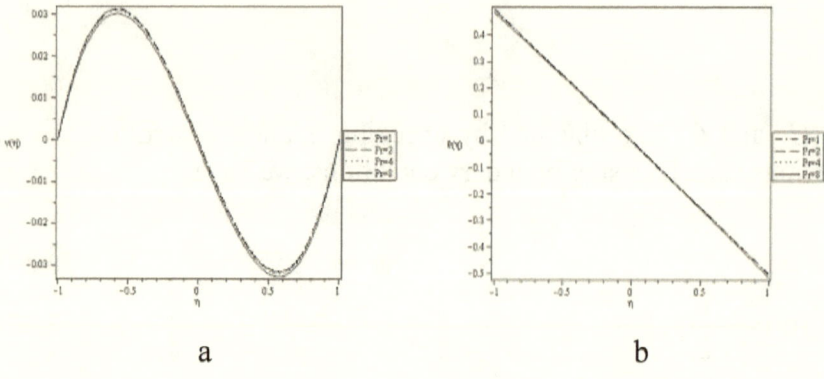

a   b

Figure 3.2 Results of various Pr when $\delta = 1, E = 1$ for a) $v(\eta)$ and b) $\theta(\eta)$

3.2 Nomenclature

A	Adomian polynomials	Pr	Prandtl number
g	a source function	E	a dimensionless number
L	linear operator	δ	a dimensionless number
N	nonlinear operator	v	dimensionless velocity
V	velocity	μ	viscosity
x	a variable	η	dimensionless velocity
y	a variable	Θ	dimensionless temperature
z	a variable	θ	temperature

3.3 References

1. A. H. Nayfeh, Perturbation Methods, Wiley, New York, 2000.
2. G. Adomian, A review of the decomposition method in applied mathematics, J. Math. Anal. Appl. 135 (1988) 501.
3. G. Adomian, A review of the decomposition method and some recent results for nonlinear equations, Math. Comput. Model, 13 (7) (1992) 17.
4. H. Jafari, V. Daftardar-Gejji, Revised Adomian decomposition method for solving a system of nonlinear equations, Appl. Math. Comput. 175 (2006) 1.
5. S. Ghosh, A. Roy, D. Roy, An adaptation of Adomian decomposition for numeric-analytic integration of strongly nonlinear and chaotic oscillators, Commun. Methods Appl. Mech. Eng. 196 (2007) 1133.
6. H. Bulut, M. Ergut, V. Asil, R. H. Bokor, Numerical solution of a viscous incompressible flow problem through an orifice by Adomian decomposition method, Appl. Math. Comput. 153 (2004) 733.
7. F. M. Allan, M. I. Syam, On the analytic solutions of the non-homogeneous Blasius problem, J. Comput. Appl. Math. 182 (2005) 362.
8. L. Wang, A new algorithm for solving classical Blasius equation, Appl. Math. Comput. 157 (2004).
9. I. Hashim, Adomian decomposition method for solving BVPs for fourth-order integro-differential equations, J. Comput. Appl. Math. 182 (2005) 362.
10. C. W. Soh, Non-perturbative semi-analytical source-type solutions of thin-film equation, J. Comput. Appl. Math. 174 (2006) 1576.
11. D. D. Ganji 1 *, F. Eskandarinia 1, Z. Z. Ganji 1, Hossain D. Ganji, Analytic solution of natural convection flow of a non-Newtonian fluid between two vertical flat plates by using decomposition method, Numerical methods in partial differential equations, DOI 10.1002/num.20584.
12. R. W. Bruce, T. Y. Na, Natural convection flow of Powell–Eyring fluids between two vertical flat plates, ASME 67 WA/HT (1967) 25.

13. A. V. Shenoy, R. A. Mashelkar, Thermal convection in non-Newtonian fluids. Advances in heat transfer, Advances in heat transfer, 15 (1982) Academic Press, New York.
14. K. R. Rajagopal, T. Y. Na, Natural convection flow of a non-Newtonian fluid between two vertical flat plates, Acta Mech 54 (1985) 239.

CHAPTER 4

4. Homotopy Analysis Method

This section is devoted to a comprehensive introduction of the Homotopy Analysis Method (HAM) (1-24). Assume the following nonlinear differential equation.

$$N[u(\tau)] = 0 \tag{4.1}$$

We define the function $\phi(\tau, p)$ as follows:

$$\lim_{p \to 0} \phi(\tau, p) = u_0(\tau) \tag{4.2}$$

where N is a nonlinear operator, τ is an independent variable, and $u(\tau)$ is the solution of equation $p \in [0,1]$, and $u_0(\tau)$ is the initial guess, which satisfies the initial or boundary conditions and

$$\lim_{p \to 1} \phi(\tau, p) = u(\tau) \tag{4.3}$$

By using the generalized homotopy method, Liao's so-called zero-order deformation equation (4.1) will be

$$(1-p)L[\phi(\tau,p)-u_0(\tau)] = p\hbar H(\tau)N[\phi(\tau,p)]$$

(4.4)

where $\hbar$ is the auxiliary parameter that helps to control the convergence of the results, $H(\tau)$ is the auxiliary function, and L is the linear operator. It should be emphasized that there one can have great freedom to choose the auxiliary parameter $\hbar$, the auxiliary function $H(\tau)$, the initial guess $u_0(\tau)$, and the auxiliary linear operator L, which is one of the main flexibilities of this method. Thus, when p increases from 0 to 1, the solution $\phi(\tau,p)$ changes between the initial guess, $u_0(\tau)$, and the solution, $u(\tau)$. One can write the Taylor series expansion of $\phi(\tau,p)$ with respect to p as

$$\phi(\tau,p) = u_0(\tau) + \sum_{m=1}^{+\infty} u_m(\tau)p^m$$

(4.5)

and

$$u_0^{[m]}(\tau) = \frac{\partial^m \phi(\tau;p)}{\partial p^m}\bigg|_{p=0}$$

(4.6)

where $u_0^{[m]}(\tau)$ is the mth order of deformation derivation which reads

$$u_m(\tau) = \frac{u_o^{[m]}}{m!} = \frac{1}{m!}\frac{\partial^m \phi(\tau;p)}{\partial p^m}\bigg|_{p=0}.$$

(4.7)

Defining the vector of

$$\vec{u}_m = \{\vec{u}_1, \vec{u}_2, \vec{u}_3, \ldots\ldots, \vec{u}_n\}$$

(4.8)

and following the definition of equation 4.7, the governing equation and the corresponding initial conditions of $u_m(\tau)$ can be deduced from zero-order deformation equation (4.1). Differentiating equation 4.1 m times with respect to the embedding parameter p and setting $p = 0$ and finally dividing by $m!$, we will have the so-called m^{th} order deformation equation in the following form:

$$L[u_m(\tau) - \chi_m u_{m-1}(\tau)] = \hbar H(\tau) R(\vec{u}_{m-1}) \qquad (4.9)$$

where

$$R_m(\vec{u}_{m-1}) = \frac{1}{(m-1)!} \frac{\partial^{m-1} N[\phi(\tau;p)]}{\partial p^{m-1}}\bigg|_{p=0} \qquad (4.10)$$

and

$$\chi_m = \begin{cases} 0 & m \leq 1 \\ 1 & m > 1 \end{cases} \qquad (4.11)$$

Finally, applying inverse linear operator to both sides of the linear equation, equation 4.1, we can easily solve the equation and compute the generation constant by applying the initial or boundary conditions.

4.1 Magneto Hydrodynamic Flow in Porous Media (20)

Assume the problem of a steady laminar boundary layer flow of an electrically conducting fluid over a porous surface (25-27). Further, assume that there is a magnetic field β, which is applied in the Y-direction, and the magnetic Reynolds number is small. Assume a negligible magnetic field and neglect the effects of viscous dissipation, Ohmic heating, and hall currents. The X-axis is set along the horizontal plate, and the Y-axis is perpendicular to it. The plate is moving with a constant speed U at a temperature T_w. The fluid for the above plate is kept at a temperature of T_∞. Under the above assumptions, the boundary layer equations governing the flow and the heat transfer over an infinite plate can be written as follows:

$$u_X + v_Y = 0,$$ (4.12)

$$uu_X + vv_Y = \xi u_{YY} - k^{-1}\xi u - cu^2 - \sigma\beta^2\rho^{-1}u,$$ (4.13)

$$uT_X + vT_Y = \rho^{-1}c_p^{-1}(Q(T - T_\infty) + K_e T_{YY}),$$ (4.14)

Here, X and Y are respectively the distances along perpendicular to the surface; u and v are the velocity components along X and Y directions respectively; and T is the temperature, ρ is the fluid density, ξ is the kinematic viscosity, c_p is the specific heat at constant pressure, K is the permeability value of porous medium, C is the Forcheimer inertia coefficient, k_e is the effective thermal conductivity, β is the magnetic induction, σ is the fluid electrical conductivity, and Q is the heat generation/absorption coefficient.

The appropriate boundary conditions are given by the following:

$$Y = 0, \quad u(X) = U, \quad v(X) = -v_w(X), \quad T(X) = T_w,$$ (4.15)

$$Y \to \infty \quad u(X) = 0, \quad T(X) = T_\infty,$$

Where U is a constant, $v_w(X) > 0$ is the fluid suction at the plate surface, and $v_w(X) < 0$ is the fluid blowing or injection at the wall. One can reduce the PDEs of equations (4.12)-(4.14) to a system of ordinary differential equations by means of similarity transformation technique. Following the definition of similarity variables,

$$Y = \left(\frac{2\xi X}{U}\right)^{0.5}\eta, \quad u = UF'(\eta),$$

$$v = (\eta F' - F)\left(\frac{\xi U}{2X}\right)^{0.5}, \quad \theta = \frac{T - T_\infty}{T_W - T_\infty},$$ (4.16)

Equation 4.12 is identically satisfied, and the partial differential equations 4.13 and 4.14 transform into following ordinary differential equations:

$$F''' + FF'' - ((M + D^{-1}) - \alpha_X F')F' = 0, \tag{4.16}$$

$$\theta'' + pr(F\theta' + Q_X\theta) = 0. \tag{4.18}$$

Which primes denote derivatives with respect to η. Free convection boundary conditions, Equation 4.15 is also transformed into the following form.

$$\begin{aligned} \eta = 0, \quad F = F_w, \quad F' = 1, \quad \theta = 1, \\ \eta \to \infty \quad F' = 0, \quad \theta = 0. \end{aligned} \tag{4.19}$$

Important dimensionless numbers used in this problem are

Hartmann number	$M = 2\sigma X \beta \left(\rho U^2\right)^{-1} (X)^2$
inverse Darcy number	$D^{-1} = 2\xi X (KU)^{-1}$
dimensionless inertia coefficient	$\alpha_x = 2CX$
Prandtl number	$pr = \rho\xi C_p K_e^{-1}$
dimensionless heat generation/ absorption coefficient	$Q_X = 2QX(U_\rho C_p)^{-1}$
dimensionless suction/blowing coefficient	$F_w = -v_w(X)\left(\dfrac{2X}{\xi U}\right)^{0.5}$

One can choose the initial guesses and auxiliary linear operator in the following forms:

$$f_0(\eta) = f_w + 1 - \exp(-\eta), \theta_0(\eta) = \exp(-\eta) \tag{4.20}$$

$$L_1(f) = f''' - f', L_2(\theta) = \theta'' - \theta \tag{4.21}$$

$$L_1(c_1 + c_2 e^{\eta} + c_3 e^{-\eta}) = 0, L_2(c_4 e^{\eta} + c_5 e^{-\eta}) = 0 \tag{4.22}$$

where $c_i (i = 1-5)$ are constants. Let $p \in [0,1]$ denote the embedding parameter and $\hbar_1, \hbar_2$ indicate the nonzero auxiliary parameters. We then construct the following problems:

Zeroth-order deformation problems

$$(1-p)L_1[f(\eta; p) - f_0(\eta)] = p\hbar_1 N_1[f(\eta; p) \theta(\eta; p)] \tag{4.23}$$

$$f(0; p) = 0, f'(0; p) = 1, f'(\infty; p) = 0 \tag{4.24}$$

$$(1-p)L_1[\theta(\eta; p) - \theta_0(\eta)] = p\hbar_2 N_2[f(\eta; p) \theta(\eta; p)] \tag{4.25}$$

$$\theta(0; p) = 1, \theta(\infty; p) = 0 \tag{4.26}$$

$$N_1[f(\eta; p), \theta(\eta; p)] = \frac{\partial^3 f(\eta; p)}{\partial \eta^3} + f(\eta; p)\frac{\partial^2 f(\eta; p)}{\partial \eta^2} -$$
$$((M + D^{-1}) - \alpha_\chi \frac{\partial f(\eta; p)}{\partial \eta})\frac{\partial f(\eta; p)}{\partial \eta} \tag{4.27}$$

$$N_2[f(\eta; p), \theta(\eta; p)] = \frac{\partial^2 \theta(\eta; p)}{\partial \eta^2} + pr(f(\eta; p)\frac{\partial \theta(\eta; p)}{\partial \eta} - Q_\chi \theta(\eta; p)) \tag{4.28}$$

For $p = 0$ and $p = 1$, we have the following:

$$f(\eta; 0) = f_0(\eta), f(\eta; 1) = f(\eta) \tag{4.29}$$

$$\theta(\eta; 0) = \theta_0(\eta), \theta(\eta; 1) = \theta(\eta) \tag{4.30}$$

When p increases from 0 to 1, then $f(\eta; p)$ and $\theta(\eta; p)$ vary from $f_0(\eta)$ and $\theta_0(\eta)$ to $f(\eta)$ and $\theta(\eta)$. Due to the Taylor series with respect to (p), we have the following:

$$f(\eta; p) = f_0(\eta) + \sum_{m=1}^{\infty} f_m(\eta) p^m , f_m(\eta) = \frac{1}{m!} \frac{\partial^m (f(\eta; p))}{\partial p^m} \qquad (4.31)$$

$$\theta(\eta; p) = \theta_0(\eta) + \sum_{m=1}^{\infty} \theta_m(\eta) p^m , \theta_m(\eta) = \frac{1}{m!} \frac{\partial^m (\theta(\eta; p))}{\partial p^m} \qquad (4.32)$$

In which $\hbar_1$ and $\hbar_2$ are chosen in such a way that these two series are convergent at ($p = 1$).

Therefore, we have through equations 4.31, 4.32 that

$$f(\eta) = f_0(\eta) + \sum_{m=1}^{\infty} f_m(\eta) , \theta(\eta) = \theta_0(\eta) + \sum_{m=1}^{\infty} \theta_m(\eta) \qquad (4.33)$$

Mth-order deformation problems

$$L_1[f_m(\eta) - \chi_m f_{m-1}(\eta)] - \hbar_1 R_m^f(\eta) \qquad (4.34)$$

$$f_m(0) = f_m'(0) = f_m'(\infty) = 0 \qquad (4.35)$$

$$L_2[\theta_m(\eta) - \chi_m \theta_{m-1}(\eta)] = \hbar_2 R_m^\theta(\eta) \qquad (4.36)$$

$$\theta_m(0) = \theta_m(\infty) = 0 \qquad (4.37)$$

$$R_m^f = f_{m-1}''' + \sum_{n=0}^{m-1} f_{m-1-n} f_n'' - (M + D^{-1}) f_{m-1}' + \alpha_\chi \sum_{n=0}^{m-1} f_{m-1-n}' f_n' \qquad (4.38)$$

$$R_m^\theta = \theta_{m-1}'' + pr(\sum_{n=0}^{m-1} f_{m-1-n} \theta_n' + Q_\chi \theta_{m-1}) \qquad (4.39)$$

$$\chi_m \begin{cases} 0 & m \le 1 \\ 1 & m > 1 \end{cases} \tag{4.40}$$

Maple 11.01 is used to solve the equations 4.23 to 4.36 up to first few orders of approximations, and it is found that f and θ can be expressed as

$$\begin{cases} f_m(\eta) = \sum_{n=0}^{2m+1} \Psi 1_{m,n}(\eta) e^{-m\eta}, & \Psi 1_{m,n}(\eta) = \sum_{n=0}^{2m+1-n} a_{m,n}^k \eta^k, \\ \theta_m(\eta) = \sum_{n=0}^{2m+1} \Psi 2_{m,n}(\eta) e^{-m\eta}, & \Psi 2_{m,n}(\eta) = \sum_{n=0}^{2m+1-n} b_{m,n}^k \eta^k, \end{cases} \tag{4.41}$$

$$\begin{cases} a_{m,0}^0 = \chi_m \chi_{2m+1} a_{m-1,0}^0 - \sum_{n=2}^{m+1}\left[(n-1)\Delta_{m,n}^0 \mu 1_{n,0}^0 + \sum_{q=1}^{2m+1-n} \Delta_{m,n}^q \left((n-1)\mu 1_{n,0}^q - \mu 1_{n,1}^q\right) \right], \\ a_{m,0}^k = \chi_m \chi_{2m+1-k} a_{m-1,0}^k, \qquad 1 \le k \le 2m+1, \\ a_{m,1}^0 = \chi_m \chi_{2m} a_{m-1,1}^0 + \sum_{q=0}^{2m} \Delta 1_{m,n}^q \mu 1_{1,1}^q + \sum_{n=2}^{2m+1}\left\{ n\Delta 1_{m,n}^0 \mu 1_{n,0}^0 + \sum_{q=1}^{2m+1-n} \Delta_{m,n}^q (n\mu 1_{n,0}^q - \mu 1_{n,1}^q) \right\}, \\ a_{m,1}^k = \chi_m \chi_{2m-k} a_{m-1,1}^k + \sum_{q=k-1}^{2m} \Delta 1_{m,1}^q \mu 1_{1,k}^q, \qquad 1 \le k \le 2m+1, \\ a_{m,n}^k = \chi_m \chi_{2m+1-n-k} a_{m-1,n}^k + \sum_{q=k}^{2m+1-n} \Delta 1_{m,n}^q \mu 1_{n,k}^q, \qquad 2 \le k \le 2m+1, \ 0 \le k \le 2m+1-n, \\ b_{m,1}^0 = \chi_m \chi_{2m} b_{m-1,1}^0 - \sum_{n=2}^{2m+1} \sum_{q=0}^{2m+1-n} \Delta 2_{m,n}^q \mu 2_{n,0}^q, \\ b_{m,1}^k = \chi_m \chi_{2m-k} b_{m-1,1}^k + \sum_{n=k-1}^{2m} \Delta 2_{m,n}^q \mu 2_{1,k}^q, \qquad 1 \le k \le 2m+1, \\ b_{m,n}^k = \chi_m \chi_{2m+1-n-q} b_{m-1,n}^k + \sum_{q=k}^{2m+1-n} \Delta 2_{m,n}^q \mu 2_{n,k}^q, \qquad 2 \le n \le 2m+1, \ 0 \le k \le 2m+1-n, \\ \mu 1_{1,k}^q = \sum_{p=0}^{q+1-k} \frac{q!}{k! 2^{q+1-k-p}}, \qquad q \ge 0, \ 1 \le k \le 2q+1, \\ \mu 1_{n,k}^q = \sum_{r=0}^{q-k} \sum_{p=0}^{q-k-r} \frac{q!}{k!(n-1)^{q+1-k-r-p} n^{r+1}(n+1)^{p+1}}, \qquad 0 \le k \le 2q+1, q \ge 0, \ n \ge 2, \\ \mu 2_{1,k}^q = \frac{q! 2^{q+2-k}}{k!}, \qquad 0 \le k \le 2q+2, \ q \ge 0, \\ \mu 2_{n,k}^q = \sum_{p=0}^{q+1-k} \frac{q!}{k!(n-1)^{p+1}(n+1)^{q+1-k-p}}, \qquad 0 \le k \le 2q+2-n, \ q \ge 0, \ n \ge 2, \end{cases} \tag{4.42}$$

$$\Delta 1_{m,n}^{q} = h_1 \left\{ \chi_m \chi_{2m+1-n-q} \left(a3_{m-1,n}^{q} - (M+D)^{-1} a1_{m-1,n}^{q} \right) + \chi_m \chi_{2m-n-q+2} (\delta 1_{m,n}^{q} + \alpha \delta 2_{m,n}^{q}) \right\};$$

$$\Delta 2_{m,n}^{q} = h_2 \left\{ \chi_m \chi_{2m+1-n-q} \left(b2_{m-1,n}^{q} + prQb_{m-1,n}^{q} \right) + \chi_m \chi_{2m-n} \delta 3_{m,n}^{q} \right\};$$

$$\delta 1_{m,n}^{q} = \sum_{k=0}^{m-1} \sum_{j=\max(0,n-2m+2k+1)}^{\min(n,2k+1)} \sum_{i=\max(0,q-2m+2k+1+n-j)}^{\min(q,2k+1-j)} a_{m-1-k,n-j}^{q-i} a2_{k,j}^{i},$$

$$\delta 2_{m,n}^{q} = \sum_{k=0}^{m-1} \sum_{j=\max(0,n-2m+2k+1)}^{\min(n,2k+1)} \sum_{i=\max(0,q-2m+2k+1+n-j)}^{\min(q,2k+1-j)} a1_{m-1-k,n-j}^{q-i} a1_{k,j}^{i},$$

$$\delta 1_{m,n}^{q} = \sum_{k=0}^{m-1} \sum_{j=\max(0,n-2m+2k+1)}^{\min(n,2k+1)} \sum_{i=\max(0,q-2m+2k+1+n-j)}^{\min(q,2k+1-j)} a_{m-1-k,n-j}^{q-i} b1_{k,j}^{i},$$

$$a1_{m,n}^{q} = (q+1)a_{m,n}^{q} - na_{m,n}^{q},$$
$$a2_{m,n}^{q} = (q+1)a1_{m,n}^{q} - na1_{m,n}^{q},$$ \qquad (4.43)
$$a3_{m,n}^{q} = (q+1)a2_{m,n}^{q} - na2_{m,n}^{q},$$
$$b1_{m,n}^{q} = (q+1)b_{m,n}^{q} - nb_{m,n}^{q},$$
$$b2_{m,n}^{q} = (q+1)b1_{m,n}^{q} - nb1_{m,n}^{q},$$

$$f_m(\eta) = \sum_{m=0}^{\infty} f_m(\eta) = \lim_{M \to \infty} \left[\sum_{m=0}^{M} a_{m,0}^{0} + \sum_{n=1}^{2M+1} e^{-m} \left(\sum_{m=n-1}^{2M} \sum_{k=0}^{2m+1-n} a_{m,n}^{k} \eta^{k} \right) \right],$$

$$\theta_m(\eta) = \sum_{m=0}^{\infty} \theta_m(\eta) = \lim_{M \to \infty} \left[\sum_{n=1}^{2M+1} e^{-m} \left(\sum_{m=n-1}^{2M} \sum_{k=0}^{2m+1-n} b_{m,n}^{k} \eta^{k} \right) \right].$$

Liao (1-6) noted that the convergence and rate of approximation for the HAM solution strongly depends on the values of auxiliary parameters $\hbar_1$ and $\hbar_2$. Figure 4 clearly depicts ranges for admissible values of $\hbar_1$ and $\hbar_2$, which are ($-2 < \hbar_1 < -0.5$) and ($-2 < \hbar 2 < -0.5$). Our calculations clearly indicate that series (22) converge for a whole region of η when $\hbar_1 = -1.2$ and $\hbar_2 = -1.4$.

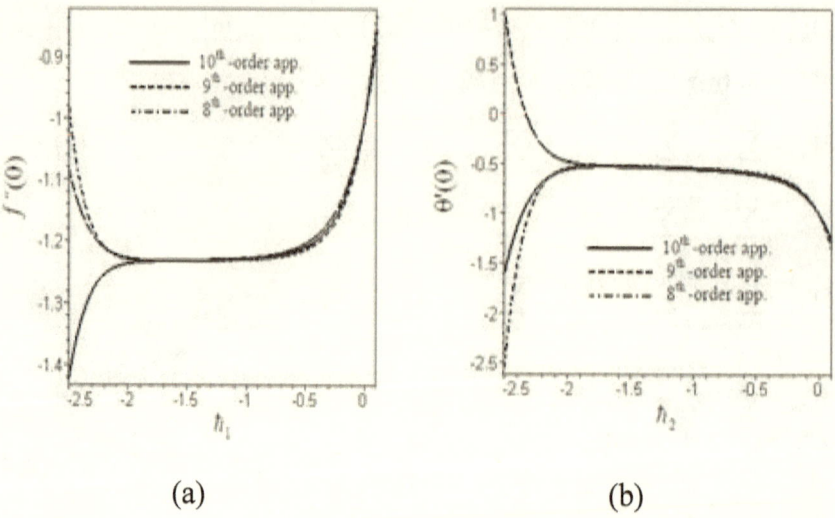

(a)   (b)

Figure 4 8th-, 9th-, and 10th-order approxiwmation assuming
$f_w = 0.1, M = 1, \alpha = 0.1, D^{-1} = 0.1$ for (a): The $\hbar_1 -$ validity and (b): The
$\hbar_2 -$ validity

The effect of surface mass transfer F_w on the dimensionless velocity
and temperature distribution is displayed in figure 3.

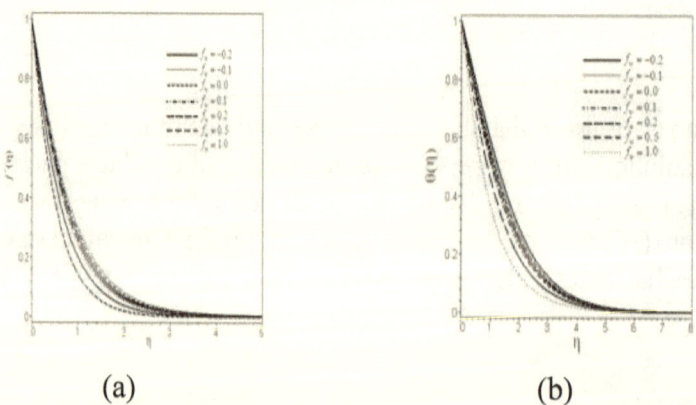

(a)   (b)

Figure 4.2 Variation of suction/blowing coefficient when

$D^{-1} = 0.1, \mathrm{Re} = 400, \alpha = 0.1, k = 0.725.$

(a) dimensionless velocity and (b) dimensionless temperature

4.2 Nomenclatures

C	Forcheimer inertia coefficient	Q	Heat coefficient
c_p	Specific heat coefficient	Y	Dimensionless length
k_e	Effective thermal conduction	θ	Dimensionless temperature
f	Dimensionless velocity	τ	An independent variable
L	Linear operator	$\hbar$	An auxiliary parameter
N	Nonlinear operator	β	Magnetic field
p	A parameter	ρ	Fluid density
u	Velocity	δ	Fluid electrical conductivity
T	Temperature		

4.3 References

1. S. J. Liao, A kind of linearity-invariance under homotopy and some simple.applications of it in mechanics, Report No. 520, Institute of Shipbuilding, University of Hamburg, 1992.
2. S. J. Liao, A new branch of solutions of boundary –layer flows over an impermeable stretched plate.Int.J.Heat Mass Transfer. 48, (2005), 2529-2539.
3. S. J. Liao, Series solutions of unsteady boundary-layer flows over a stretching flat plate, Stud.App. Math. 17, (2006) 239-264.
4. S. J. Liao, E.Magyari, Exponentially decaying boundary-layers as limiting cases of families of algebraically decaying ones. Z. Angew.Math.Phys. 57, (2006)777-792.
5. S. J. Liao, The proposed homotopy analysis technique for the solution of nonlinear problems, PhD thesis, Shanghai Jiao Tong University, 1992.
6. S. J. Liao, Beyond Perturbation: Introduction to the Homotopy Analysis Method, Chapman & Hall/CRC Press, Boca Raton, 2003.
7. Z. K. Wang, T. A. Cao, an Introduction to Homotopy Methods, Chongqing Publishing House, Chongqing, 1991.
8. C. Nash, S. Sen, Topology and Geometry for Physicists, Academic Press, Inc, London, 1983.

9. J. M. Ortega, W. C. Rheinboldt, Iterative solution of non-linear equations in several variables, Academic press, New York, 1970.

10. T.Hayat, M.Khan, S.Asghar, Magneto hydrodynamic flow of an oldroyd 6-constant fluid.Appl. Math.Compute. 155(2004)417-25.

11. T.Hayat, M.Sajid, An analytic solution for thin film flow of a forth grade fluid down a vertical cylinder.Phys.Lett A. 361(2007)316-322.

12. T.Hayat, M.Khan, Homotopy solutions for a generalized second grade fluid past a porous plate.Nonlinear Dyn. 24, (2005)395-405.

13. T.Hayat, M.Khan, M.Ayub, on non-linear flows with slip boundary condition, Z.Angew.Math.Phys.56, (2005)1012-1029.

14. M.Sajid, T.Hayat, S.Asghar, on the analytic solution of the steady flow of a forth grade fluid.Phys.Lett A. 355, (2006)18-26.

15. Z.Abbas, M.Sajid, T.Hayat, MHD boundary layer flow of an upper -convected Maxwell fluid in a porous channel.Theor. Comput.Fluid Dyn.20, (2006)229-238.

16. T.Hayat, Z.Abbas, M.Sajid, S.Asghar, The influence of thermal radiation on MHD flow of a second grade fluid.Int.J.Heat Mass Transfer.50, (2007)931-941.

17. S.Abbasbandy, The application of homotopy analysis method to nonlinear equations arising in heat transfer.Phys.Lett A. (360), (2006)109-113.

18. S.Abbasbandy, Homotopy analysis method for heat radiation equations.Int.Commun.Heat and Mass Transfer. (34), (2007)380-387.

19. S. Abbasbandy.Approximate solution for the nonlinear model of diffusion and reaction in porous catalysts by means of the homotopy analysis method Chemical Engineering Journal.136 (2-3), (2008)144-150.

20. Z. Ziabakhsh, G. Domairry.Solution of the laminar viscous flow in a semi-porous channel in the presence of a uniform magnetic field by using the homotopy analysis method.Communications in Nonlinear Science and Numerical Simulation, In Press.

21. G. Domairry, N. Nadim.Assessment of homotopy analysis method and homotopy perturbation method in non-linear

heat-transfer equation International Communications in Heat and Mass Transfer. 35(1) (2008) 93-102.

22. M. M. Rashidi, G. Domairry, S. Dinarvand.Approximate solutions for the Burger and regularized long wave equations by means of the homotopy analysis method Communications in Nonlinear Science and Numerical Simulation, In Press.

23. G. Domairry, M. Fazeli.Homotopy analysis method to determine the fin efficiency of convective straight fins with temperature-dependent thermal conductivity Communications in Nonlinear Science and Numerical Simulation, In Press.

24. D. G. Domairry, A. Mohsenzadeh, M. Famouri.The application of homotopy analysis method to solve nonlinear differential equation governing Jeffery–Hamel flow Communications in Nonlinear Science and Numerical Simulation, In Press.

25. W. M. Kays, M. E. Crawford, Convection heat and mass transfer, Mc Craw-Hill, New York, 1993.

26. H.Schilichtinh, Boundary layer theory, 6th Edn, McGraw-Hill, New York. (1998).

27. Donald A.Nield, Adrian Bejan, Convection in porous media, 3th Edn, Springer, (2006).

CHAPTER 5

5. Differential Transformation Method

In this part, a fundamental basic of the Differential Transformation Method (1-4) is presented. A Fin problem is considered in the next section, to apply this method on that. To understand the concept of the method of Differential transformation, suppose $x(t)$ to be an analytic function in a domain D, and $t = t_i$ represents any point in D. The function $x(t)$ is then represented by one power series whose center is located at t_i. The Taylor series expansion function of $x(t)$ is of the form (5, 6)

$$x(t) = \sum_{k=0}^{\infty} \frac{(t - t_i)^k}{k!} \left[\frac{d^k x(t)}{dt^k} \right]_{t=t_i} \qquad \forall t \in D \cdot \qquad (5.1)$$

The Maclaurin series of $x(t)$ can be obtained by taking $t_i = 0$ in equation 5.1, expressed as

$$x(t) = \sum_{k=0}^{\infty} \frac{t^k}{k!} \left[\frac{d^k x(t)}{dt^k} \right]_{t=0} \qquad \forall t \in D \cdot \qquad (5.2)$$

As explained in 7, the differential transformation of the function $x(t)$ is defined as follows:

46

$$X(k) = \sum_{k=0}^{\infty} \frac{H^k}{k!} \left[\frac{d^k x(t)}{dt^k} \right]_{t=0}, \qquad (5.3)$$

Where $x(t)$ and $X(k)$ are the original function and the transformed function respectively. The differential spectrum of $X(k)$ is confined within the interval $t \in [0, H]$, where H is a constant value. The differential inverse transform of $X(k)$ is defined as follows:

$$x(t) = \sum_{k=0}^{\infty} \left(\frac{t}{H} \right)^k X(k) \qquad (5.4)$$

As mentioned before, the concept of differential transformation is based upon the Taylor series expansion. The values of function $X(k)$ at values of argument k are referred to as *discrete*, i.e., $X(0)$ is known as the zero discrete, $X(1)$ as the first discrete, etc. The more discretes available, the more precise it is possible to restore the unknown function. The function $x(t)$ consists of the T-function $X(k)$, and its value is given by the sum of the T-function with $(t/H)^k$ as its coefficient. The function $x(t)$ is expressed by a finite series, and equation 5.4 can be written as

$$x(t) = \sum_{k=0}^{n} \left(\frac{t}{H} \right)^k X(k). \qquad (5.5)$$

A list of mathematical operations used in the differential transform method is listed in table 5.1

Table 5.1 The fundamental operations of the differential transform method

Original function	Transformed function
$x(t) = \alpha f(x) \pm \beta g(t)$	$X(k) = \alpha F(k) \pm \beta G(k)$
$x(t) = \dfrac{df(t)}{dt}$	$X(k) = (k+1)F(k+1)$
$x(t) = \dfrac{d^2 f(t)}{dt^2}$	$X(k) = (k+1)(k+2)F(k+2)$
$x(t) = \exp(t)$	$X(k) = \dfrac{k}{k!}$
$x(t) = f(t)g(t)$	$X(k) = \displaystyle\sum_{l=0}^{k} F(l)G(k-l)$
$x(t) = t^m$	$X(k) = \delta(k-m) = \begin{cases} 1 & k = m \\ 0 & k \neq m \end{cases}$

5.1 Convective Straight Fin (8)

Analyzing the fin problem (9, 10) consists of two sections. One is based on thermal convection through fin surface area where the heat-transfer rate of fin is a product of three factors: heat-transfer coefficients, fin surface area, and temperature difference between fin surface and its surroundings. The second one is based on thermal conduction through fin cross section, where the main factors that govern the heat removal of fin are thermal conductivity, fin cross-section area, and temperature gradient along heat flow direction. Recently, this kind of problem has been solved and analyzed by some researchers using different methods (11-21).

Consider a straight fin with a temperature-dependent thermal conductivity, arbitrary constant cross-sectional area A_c, perimeter P, and length b (see figure 5.1). The fin is attached to a base surface of temperature T_b, it extends into a fluid of temperature T_a, and we assume that the tip of the fin is insulated.

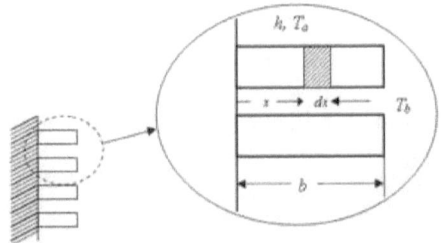

Figure 5.1 The sketch of the problem under discussion

The one-dimensional energy balance equation governing the fin problem is

$$A_c \frac{d}{dx}\left[k(T)\frac{dT}{dx}\right] - Ph(T_b - T_a) = 0 \cdot \qquad (5.6)$$

Assume a linear function of temperature for the thermal conductivity of the fin material,

$$k(T) = k_a\left[1 + \lambda(T - T_a)\right], \qquad (5.7)$$

where k_a is the thermal conductivity at the ambient fluid temperature of the fin, and k is the parameter describing the thermal conductivity variation.

Employing the following dimensionless parameters (12)

$$\theta = \frac{T - T_a}{T_b - T_a}, \quad \zeta = \frac{x}{b}, \quad \beta = \lambda(T_b - T_a), \quad \psi = (\frac{hPb^2}{k_a A_c})^{1/2}. \qquad (5.8)$$

The governing equation reduces to

$$\frac{d^2\theta}{d^2\zeta} + \beta\theta\frac{d^2\theta}{d^2\zeta} + \beta(\frac{d\theta}{d\zeta})^2 - \psi^2\theta = 0, \qquad (5.9)$$

$$\frac{d\theta}{d\zeta} = 0 \quad \text{at} \ \zeta = 0, \qquad (5.10)$$

$\theta = 1$ at $\zeta = 1$. $\qquad\qquad$ (5.11)

According to Newton's law of cooling, the heat-transfer rate from the fin is expressed as

$$Q = \int_0^b P(T - T_a)dx \, . \qquad\qquad (5.12)$$

The fin efficiency is the ratio of actual heat transfer from the fin surface to the other side while the whole fin surface is at the same temperature, described as follows:

$$\eta = \frac{Q}{Q_{ideal}} = \frac{\displaystyle\int_0^b P(T - T_a)dx}{Pb(T_b - T_a)} = \int_{\zeta=0}^{1} \theta(\zeta)d\zeta \qquad\qquad (5.13)$$

Taking the differential transform of equation 5.9 with respect to y and considering $H = 1$ gives us the following:

$$(k+1)(k+2)\Theta(k+2) + \beta \sum_{i=0}^{k}[\Theta(i)(k-m+1)(k-m+2)\Theta(k-m+2) \qquad (5.14)$$
$$+(i+1)\Theta(i+1)(k-m+1)\Theta(k-m+1)] - \psi^2\Theta(k) = 0.$$

Using boundary conditions, equation 5.10 at $y = 0$, and applying the transformation, we have

$$\Theta(1) = 0, \qquad\qquad (5.15)$$

and for the second B.C., we have

$$\Theta(0) = a \qquad\qquad (5.16)$$

where a is constant, which will be calculated later.

$$\Theta(2) = \frac{1}{2}\frac{\psi^2 a}{1+\beta a}$$

$$\Theta(3) = 0$$

$$\Theta(4) = -\frac{1}{24}\frac{\psi^4 a(2\beta a - 1)}{(1+\beta a)^3}$$

$$\Theta(5) = 0$$

$$\Theta(6) = \frac{1}{720}\frac{\psi^6 a(2\beta a - 1)(14\beta a - 1)}{(1+\beta a)^5}$$

$$\Theta(7) = 0$$

$$\Theta(8) = -\frac{1}{40320}\frac{\psi^8 a(2\beta a - 1)(25592\beta^3 a^3 - 7152\beta^2 a^2 + 330\beta a - 1)}{(1+\beta a)^7}$$

$$\Theta(9) = 0$$

$$\vdots$$

$$(5.17)$$

This is a successive process. Substituting equation 5.17 into the main equation based on DTM, it can be obtained that the closed form of the solutions is as follows:

$$\theta(\zeta) = a + \frac{1}{2}\frac{\psi^2 a}{1+\beta a}\zeta^2 - \frac{1}{24}\frac{\psi^4 a(2\beta a - 1)}{(1+\beta a)^3}\zeta^4 + \frac{1}{720}\frac{\psi^6 a(2\beta a - 1)(14\beta a - 1)}{(1+\beta a)^5}\zeta^6$$
$$- \frac{1}{40320}\frac{\psi^8 a(2\beta a - 1)(25592\beta^3 a^3 - 7152\beta^2 a^2 + 330\beta a - 1)}{(1+\beta a)^7}\zeta^8 + \dots$$

$$(5.18)$$

To obtain the value of a, we substitute the boundary condition from equation 5.11 into equation 5.18 in point $y = 1$. So we have

$$\theta(1) = a + \frac{1}{2}\frac{\psi^2 a}{1+\beta a} - \frac{1}{24}\frac{\psi^4 a(2\beta a - 1)}{(1+\beta a)^3} + \frac{1}{720}\frac{\psi^6 a(2\beta a - 1)(14\beta a - 1)}{(1+\beta a)^5}$$
$$- \frac{1}{40320}\frac{\psi^8 a(2\beta a - 1)(25592\beta^3 a^3 - 7152\beta^2 a^2 + 330\beta a - 1)}{(1+\beta a)^7} + \dots = 1$$

$$(5.19)$$

Solving equation 5.19 gives the value of a. By substituting obtained a into equation 5.18, we can find the expressions of $\theta(\zeta)$. For the case of constant thermal conductivity ($\beta = 0$), results of the present analysis are tabulated against the analytical solution (11) in table 5.2.

Davood Domairry Ganji & Ehsan Mohseni Languri

Table 5.2 The results of DTM and exact solution for $\theta(\xi)$

ξ	$\beta = 0, \psi = 0.5$			$\beta = 0, \psi = 1$		
	DTM	Exact (10)	Error	DTM	Exact (10)	Error
0	0.8868188841	0.8868188838	0.0000000003	0.6480542737	0.6480542738	0.0000000001
0.05	0.8870960292	0.8870960294	0.0000000002	0.6488645100	0.6488645103	0.0000000003
0.10	0.8879276383	0.8879276382	0.0000000003	0.6512972462	0.6512972462	0.0000000000
0.15	0.8893142310	0.8893142313	0.0000000003	0.6553585647	0.6553585646	0.0000000001
0.20	0.8912566748	0.8912566745	0.0000000003	0.6610586207	0.6610586205	0.0000000002
0.25	0.8937561827	0.8937561822	0.0000000005	0.6684116668	0.6684116676	0.0000000008
0.30	0.8968143173	0.8968143168	0.0000000005	0.6774360915	0.6774360918	0.0000000003
0.35	0.9004329897	0.9004329893	0.0000000004	0.6881544588	0.6881544591	0.0000000003
0.40	0.9046144623	0.9046144616	0.0000000007	0.7005935709	0.7005935709	0.0000000000
0.45	0.9093613475	0.9093613475	0.0000000000	0.7147845318	0.7147845315	0.0000000003
0.50	0.9146766135	0.9146766141	0.0000000006	0.7307628258	0.7307628261	0.0000000003
0.55	0.9205635842	0.9205635830	0.0000000012	0.7485684079	0.7485684081	0.0000000002
0.60	0.9270259345	0.9270259345	0.0000000000	0.7682458015	0.7682458010	0.0000000005
0.65	0.9340677074	0.9340677069	0.0000000005	0.7898442081	0.7898442088	0.0000000007
0.70	0.9416933025	0.9416933025	0.0000000000	0.8134176386	0.8134176383	0.0000000003
0.75	0.9499074872	0.9499074868	0.0000000004	0.8390250359	0.8390250359	0.0000000000
0.80	0.9587153946	0.9587153943	0.0000000003	0.8667304332	0.8667304328	0.0000000004
0.85	0.9681225298	0.9681225300	0.0000000002	0.8966031067	0.8966031075	0.0000000008
0.90	0.9781347745	0.9781347735	0.000000001	0.9287177570	0.9287177568	0.0000000002
0.95	0.9887583840	0.9887583835	0.0000000005	0.9631546839	0.9631546843	0.0000000004
1.00	1.000000000	0.9999999999	0.0000000001	1.000000001	1.000000000	0.000000001

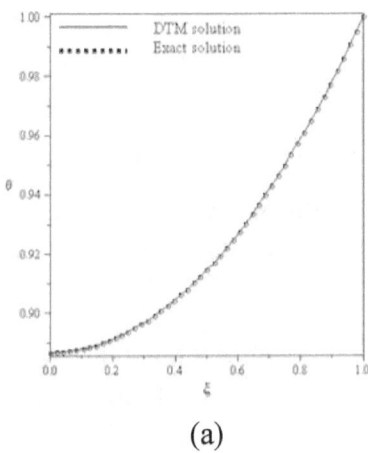
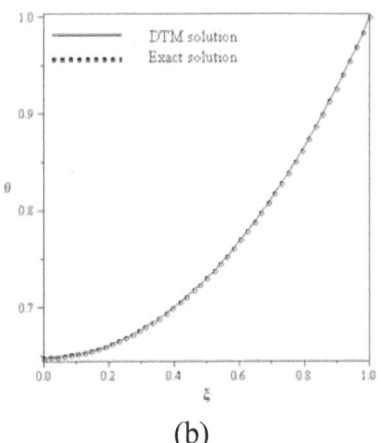

(a) (b)

Figure 5.2 Comparison of the solutions via DTM and exact solution for $\theta(\xi)$ (a): $\beta = 0$, $\psi = 0.5$ (b): $\beta = 0$, $\psi = 1$

5.2 Nomenclature

A_c	cross-sectional area of the fin (m^2)
b	fin length (m)
DTM	Differential Transformation Method
h	heat-transfer coefficient ($W\ m^{-1}\ K^{-1}$)
k	thermal conductivity of the fin material ($W\ m^{-1}\ K^{-1}$)
k_a	thermal conductivity at the ambient fluid temperature ($W\ m^{-1}\ K^{-1}$)
k_b	thermal conductivity at the base temperature ($W\ m^{-1}\ K^{-1}$)
NS	Numerical Solution
P	fin perimeter (m)
Q	heat-transfer rate (W)
T_a	temperature of surface a (K)
T_b	temperature of surface b (K)
x	distance measured from the fin tip (m)
β	dimensionless parameter describing variation of the thermal conductivity
η	fin efficiency
ζ	dimensionless coordinate
λ	the slope of the thermal conductivity temperature curve (K^{-1})
ψ	thermogeometric fin parameter
θ	dimensionless temperature

5.3 References

1. J. K. Zhou, Differential Transformation and its Applications for Electrical Circuits, Huarjung University Press, Wuuhahn, China, (1986) (in Chinese).
2. J. S. Chiou, J. R. Tzeng, Application of the Taylor transform to nonlinear vibration problems, Transaction of the American Society of Mechanical Engineers, Journal of Vibration and Acoustics, 118: 83-87 (1996).
3. M. J. Jang, C. L. Chen, Y. C. Liu, Two-dimensional differential transform for partial differential equations, Appl. Math. Comput. 121: 261-270 (2001).
4. I. H. Abdel-Halim Hassan, On solving some eigenvalue-problems by using a differential transformation, Appl. Math. Comput. 127: 1-22 (2002).
5. C. K. Chen, S. P. Ju, Application of differential transformation to transient advective–dispersive transport equation, Appl. Math. Comput. 155: 25-38 (2004).
6. C. K. Chen, S. S. Chen, Application of the differential transformation method to a non-linear conservative system, Appl. Math. Comput. 154: 431-441(2004).
7. J. K. Zhou, Differential Transformation and its Applications for Electrical Circuits, Huarjung University Press, Wuuhahn, China, (1986) (in Chinese).
8. A. A. Joneidi, D. D. Ganji, M. Babaelahi, differential transformation method to determne fin efficiency of convective straight fins with temperature dependent thermal conductivity, International communications in heat and mass transfer, 36 (2009) 757-762.
9. D. Q. Kern, D. A. Kraus, Extended Surface Heat Transfer, McGraw-Hill, New York, (1972).
10. Kraus A, Aziz A, Welty J Extended surface heat transfer. Wiley, New York, (2001).
11. Cihat Arslanturk, A decomposition method for fin efficiency of convective straight fins with temperature-dependent thermal conductivity, International Communications in Heat and Mass Transfer 32 (2005) 831-841.
12. G. Domairry, M. Fazeli, Homotopy analysis method to determine the fin efficiency of convective straight fins with

temperature-dependent thermal conductivity, Communications in Nonlinear Science and Numerical Simulation. doi:10.1016/j.cnsns.2007.09.007

13. S. M. Zubair, A. Z. Al-Garni, J. S. Nizami, The optimal dimensions of circular fins with variable profile and temperature-dependent thermal conductivity, Int. J. Heat Mass Transfer 39 (16) 3431-3439 (1996).

14. EMA Mokheimer, Performance of annular fins with different profiles subject to variable heat transfer coefficient, Int. J. Heat Mass Transfer 45 3631- 3642 (2002).

15. H. S. Kou, J. J. Lee, C. Y. Lai, Thermal analysis of a longitudinal fin with variable thermal properties by recursive formulation, Int. J. Heat Mass Transfer 48 2266-2277 (2005).

16. Aziz A, Enamul Hug SM, Perturbation solution for convecting fin with variable thermal conductivity. ASME J Heat Transf 97C 300-301 (1975).

17. Cihat Arslanturk, Correlation equations for optimum design of annular fins with temperature dependent thermal conductivity. Heat Mass Transfer in press. Doi: 10.1007/s00231-008-0446-9.

18. Shadi Mahjoob, Kambiz Vafai, A synthesis of fluid and thermal transport models for metal foam heat exchangers. International Journal of Heat and Mass Transfer, 51 15-16, (2008) 3701-3711.

19. Kambiz Vafai, Lu Zhu, Analysis of two-layered micro-channel heat sink concept in electronic cooling, International Journal of Heat and Mass Transfer, 42 12 (1999), 2287-2297.

20. Dae-Young Lee, Kambiz Vafai, Analytical characterization and conceptual assessment of solid and fluid temperature differentials in porous media, International Journal of Heat and Mass Transfer, 42 3 (1999), 423-435.

21. K. Hooman, H. Gurgenci, A. A. Merrikh, Heat transfer and entropy generation optimization of forced convection in porous-saturated ducts of rectangular cross-section, International Journal of Heat and Mass Transfer, 50 11-12 (2007) 2051-2059.

www.ingramcontent.com/pod-product-compliance
Lightning Source LLC
Chambersburg PA
CBHW021926170526
45157CB00005B/2207

9 781456 808280